THE

ANNUAL ADDRESS

BEFORE THE

HARVARD NATURAL HISTORY SOCIETY,

DELIVERED BY

REV. THOMAS HILL.

Thursday, May 19, 1853.

CAMBRIDGE:
JOHN BARTLETT.
1853.

THE

ANNUAL ADDRESS

BEFORE THE

HARVARD NATURAL HISTORY SOCIETY,

DELIVERED BY

REV. THOMAS HILL.

Thursday, May 19, 1853.

CAMBRIDGE:
JOHN BARTLETT.
1853.

CAMBRIDGE:
METCALF AND COMPANY, PRINTERS TO THE UNIVERSITY.

CAMBRIDGE, May 20th, 1853.

DEAR SIR :—

WE have the honor of informing you that the Natural History Society unanimously voted that a copy of your highly interesting and instructive Address be requested for publication.

The Address not only defined clearly and satisfactorily the true place and objects of the science of Natural History, and defended the study of nature from the imputations which are sometimes cast upon it, but also treated with great soundness and liberality of many extremely important questions of general interest.

Deeming your remarks upon these subjects worthy of a more careful consideration than could be given at a single hearing, we trust that you will comply with the request of the Society.

Most respectfully, your obedient servants,

ARTHUR T. LYMAN, Pres.
CHARLES W. ELIOT,
ALFRED HOSMER,
JOSHUA KENDALL, } *Committee.*

REV. THOMAS HILL.

WALTHAM, May 28th, 1853.

GENTLEMEN :—

IN compliance with your very kind invitation, I herewith send the manuscript of my Address on the 19th instant, only regretting that I have not leisure to elaborate more fully the closing part of the lecture, to make it more worthy of the subject and of the attention of your Society.

Yours, very truly,

THOMAS HILL.

MR. ARTHUR T. LYMAN, AND OTHERS.

ADDRESS.

The musician taking his place in a vast orchestra, and putting forth his energies upon his own single instrument, is not perhaps so well qualified to judge of the effect of the music, as one who is seated at a convenient distance, where he hears all the instruments equally well, and none of them is, through its disproportionate want of distance, disproportionately loud. Nor is it necessary that the critic should be himself a proficient upon any instrument; this might indeed withdraw his attention from the harmony, to a criticism upon that particular part of the orchestra. It is only necessary that he should have a good ear, and a large and susceptible heart, capable of feeling and of telling what it feels.

In like manner, we find that many of the best general reviews of science are by those who themselves take no active part in the pursuit of any branch. Lord Bacon has by his general views, without any real scientific power or attainment, as truly served mankind, and done as real service to the cause of Science, as any one of those who have labored for her more directly and made a thousand-fold more progress in actual knowledge. And what is done among the great men of earth is but a more shining example of that which is done daily among men of less ability. I trust these remarks may prove an acceptable apology for one who makes no pretension to the rank of a naturalist having the boldness to accept an invitation from the Harvard Society of Natural History, and to attempt a discussion of the field, spirit, and methods of that circle of sciences.

In its broadest sense Natural History would comprise the whole of creation, completely filling one of the five great departments of human knowledge. For in my view of the field of human knowledge I suppose that every thing existing in the present, past, or future may be embraced under the five great heads of Theology, Psychology, History, Natural History, and Mathematics.

In Comte's great work on Positive Philosophy, he excludes the first two of these, Theology and Psychology, declaring them to be wholly vain and worthless. I think, however, that this arose more from his prejudices than his logic, and that in fact his own system or hierarchy of sciences is incomplete without these two. In order that you may the more fully enter into the spirit of my remarks upon Natural History, allow me to expand somewhat this general view of all possible existences.

I begin, then, with Theology. This looks to the absolute origin of things, the Eternal Will which called all things into being, the Eternal Wisdom which guides all things by law, the Eternal Love which makes all things work together for good. Its field is the absolute Primal Cause, God the Infinite Spirit, the uncreated Creator.

The second human science is Psychology. This has regard to man in his essence, in his free spirit; man as revealed to himself by consciousness; man as a demigod, a subordinate cause of effects, designer of purposes, executor of plans; a finite spirit, a created creator.

Thirdly we have History, a view of the deeds of man, of the creations of the finite mind.

Fourthly comes Natural History, treating of the acts of God, of the creations of the Infinite Spirit.

Fifthly, there remains the science of Space and Time, a consideration of the field wherein these histories have past and are passing; a field to be regarded as neither created nor creating.

In the study of these five sciences, from the Mathematics to Theology, we everywhere build upon two sorts of data, one derived from consciousness, the other from observation. In the Mathematics our ideas of Time come from consciousness, of Space from observation. In Natural History the conception of force, of cause, of life, is from within us, the phenomena are perceived by sense. In History, we in like manner judge of causes and motives from consciousness, but learn the facts from testimony. And in Psychology and Theology, although the intuitions are the chief sources of knowledge, they are nevertheless guided and modified by our previous acquaintance with History and Natural History.

The five sciences are naturally arranged in this order, in which those least dependent on consciousness are placed first, and those least dependent on observation last.

This is the logical order in which they are to be pursued by the human mind. Geometry is the first lesson of the infant, and although strangely and foolishly neglected in our common-school education, it is diligently studied, even by the most unlearned, from the diagrams of creation. It is of all studies least dependent on intuitions, and therefore best adapted for the undeveloped mind. But Geometry, and the Mathematics in general, have a higher claim to be the first study of a scholar, than their mere adaptation to the weakness of the inexperienced student. They are the necessary prelude to the understanding of other sciences. Natural History, in the wide sense in which I use the word, treats of the phenomena of creation, phenomena manifested in Space and Time, and therefore inexplicable to one wholly ignorant of the laws of Space and Time. A general knowledge of the principles of Geometry and Arithmetic is absolutely essential to any knowledge of material things. And the more minute, thorough, and systematic the acquaintance with mathematical principles, the better prepared is one for the study of Physics. Almost all crude and irrational theories of physics, broached or maintained in the world, may be decisively overturned by the simple application of geometric or algebraic calculus to them.

Natural History is the second of the great divisions of human science. It takes logical precedence of History, because man acts upon the material which his Creator has furnished for him, and the knowledge of this material is indispensable to the understanding of man's action upon it. A second reason why it takes precedence of History is, that it is more dependent upon observation, less upon consciousness. It belongs, therefore, to a less mature state of mind. I do not herein speak disrespectfully of the science. The Mathematics give full exercise to the brightest minds of the human family, yet I pronounce them fitted for the child's first studies. Natural History is in like manner a full task for the noblest minds on earth, yet I consider it as properly the second study in the progress of human knowledge.

Before I enter more fully upon the subject, let me finish this rapid survey of the hierarchy of science.

We have, then, in the third place, History; under which general head I comprise the whole record of human thought and action. I have already shown why this must follow after Mathematics and Natural Science. That it must precede Psychology and Theology is shown, both from the fact that it leans more upon observation and less upon consciousness than they, and also from the fact that the actions and thoughts of men are the great indices of their souls, and therefore History in this wide sense becomes the natural mother of Psychology.

This fourth division of human knowledge is utterly rejected by Comte, as being incapable of reduction to a positive scientific form. Let us pause, therefore, for a moment, to see whether it rests upon entirely false foundations.

All the sciences comprised under the three grand divisions of which I have already spoken are admitted by Comte into his hierarchy of positive sciences, because they rest upon observation. But they rest also upon a transcendental basis. Time is not an element of observation, but of intuition. None of the senses take cognizance of duration. Touch gives us extent, and sight gives us direction, and all the senses combine to reveal to us the various phenomena of motion, such as contact, taste, sound, and color. But in motion time is hidden, not revealed; and the element of duration is eliminated by an introversion of the mind, essentially different in its nature from that effort of conception by which form is projected upon the imaginary diagrams of the geometer. I maintain, therefore, that the very Mathematics themselves rest half upon the soul, and but half upon the world.

And much more do we in Natural History build upon consciousness. To come at once to the highest point, the observation of the moral and intellectual phenomena of man; it is more than error, it is downright absurdity, for Comte to talk of reducing all mental science to phrenological physiology. Granting with him the fundamental truthfulness of Dr. Gall's method, nay, acknowledging, if you please, the claims of the latest charlatans in their so-called science of Phrenology, it would still remain true that consciousness is the main source of our knowledge upon intellectual and moral questions. For our observation of organic developments, and their comparison with known characters, can be interpreted only by our consciousness of similar elements in ourselves. The blind man cannot conceive of colors; no more easily can one without an artist eye conceive of the inspiration of a Titian. It is only from consciousness, whose testimony Mill and Comte profess to reject, that they can obtain any intelligible idea of the intellectual and moral processes of which they write.

When, therefore, we have pushed our inquiries into the organization of man, and his actions, to the utmost, so that, if you please, we have a perfect knowledge of the connection between his actions and the modifications of his cerebral and nervous system, it must be still true that we can appeal only to consciousness to explain the connection between his acts and his thoughts. Here is the limit of Comte's Positive Philosophy. Professing to rest only on observation, it cannot go behind the phenomena in any instance. It cannot prove the existence of any human thought or passion, but must deal only with men's acts. The very act of uttering his philosophy, as if to men capable of thinking, implied that he had, in his practical good sense, transcended his self-imposed limits.

We are then obliged to have a science higher than that of the brain and nervous system, a science of the interior thoughts, the powers and passions of the man. It builds partly upon the observations of phrenological physiology, but chiefly on consciousness, the only possible interpreter of those observations; which adds much more than is contained in the facts it interprets.

And, fifthly, we rise to Theology. For as we have shown that consciousness is a necessary basis of science, on which even Mill and Comte unwittingly build, to consciousness we appeal.

And the first and most important testimony of the spirit itself to its own nature is the assertion that we are free beings; having knowledge, as none can deny; having within the limits of our knowledge power, as none can deny; and having within the limits of that power freedom of will, to do or to forbear, to cause or to refrain. They who deny this testimony of consciousness are so few in number, that their denial of it may safely be attributed to some unhealthy peculiarity of mental constitution.

We are conscious of power, we know that we produce effects, — that by an act of will we cause motion, and fulfil our purposes; and we derive from undoubted induction the law of the inertia of matter. The existence of an Infinite Will and an Eternal Wisdom is the very first conclusion drawn from this double datum, the consciousness of power and will within, the law of inertia without. Thus Theology is the last of the five sciences; but it must arise from Psychology, and that must spring from History, as truly as History follows Natural History, or Natural Science follows the Mathematics.

And the Creator to whom we are thus led by Science agrees with no god of the Heathen pantheon, and with no abstraction of the dreamy poet. He is such as the Hebrew lawgiver proclaimed to the race of Israel; such as the Apostle announced on the Areopagus. He is the Lord that doeth whatsoever he pleaseth in heaven and upon earth. The argument of Paul upon his quotation of Aratus, —

τοῦ γὰρ καὶ γένος ἐσμέν,

is as sound against some modern Christians, as it was then against idolaters. For since consciousness declares our freedom and power, we ought not to think our Creator is circumscribed in his liberty, or fettered in his power. The single and sure step by which we arrive at the existence of this Being gives us a Being who possesses, in unlimited extent, whatsoever spiritual powers we ourselves have. The perverted ingenuity by which Mr. Mill in his system of logic compares this argument to that by which one might prove there was pepper in a cook by showing pepper in the broth, must have been, even to himself, more amusing than convincing. The cases would have been more nearly parallel had he argued that the cook had pepper; as no one claims that the cook created the ingredients of the broth.

I do not think that any man, whose acquaintance with Science is at all intimate, will be led by her to any such denial of the power of God; any denial of the fact that we are made in his image, that we rise to a knowledge of him through a knowledge of ourselves, and that whatever powers we possess, he has in an unlimited degree.

There has been, ever since the revival of letters, a species of controversy going on among the believers in Christ, in which some disbelievers also have joined, concerning the relations of Science and the Bible. I cannot but think that this controversy is as useless as it has been to me distasteful. The Bible is not a text-book in Science, but the records of two revelations to man.

The scientific man, in his pursuit of Science, is to have no reference to his Bible. If he is wise, he will take it as the chief guide in his moral and religious duties; but if he is wise, he will likewise carefully avoid allowing it to influence his scientific opinions. The student of the Bible must seek in it for spiritual light and life, and must interpret it by the light, not of modern science, but of the science of the day in which it was written. The truth derived, by the special student of Science and the special student of Scripture, from these two independent sources, will harmonize.

If the Mosaic account of creation appears to clash with scientific truth, it seems to me more reasonable to investigate the real connection between that account and the lawgiving in Sinai, than to endeavor to force the language of Genesis and the facts of Geology into temporary agreement.

As for the bearing of Natural Science upon the question of the New Testament, it is purely imaginary. No researches of the present or of the future can give any testimony beyond that which our fathers had to the sanctity and inviolability of natural laws. It is a childish and weak argument to bring against belief in Jesus's power to suspend natural laws, to say that these laws are not now suspended. The question is one to be settled by testimony. Were they suspended by Jesus? A modern school of critics declares that no testimony is sufficient to prove the fact of a miracle; because, they say, God being one, his work is one, and he can act upon his universe only as a whole, only as a unit. He cannot, therefore, exert a special providence, or work a special miracle. But it seems to me that this argument proves too much, and places the believer in it at once on the side of the complete and thorough atheist. It sweeps away, at a breath, all the foundations of Psychology and Theology. For if we can trust in any intuitions of our own mind, or in any deductions from the data of consciousness, we are ourselves also units. Yet we are not incapable of special action. We are free to act in each particular case, and we cannot believe God is fettered.

Moreover, the testimony of the very sciences which are quoted against miracles is in their favor. For unless we accept the fanciful speculations of Lamarck, and Demaillet, woven into such popular form in the book entitled "Vestiges of Creation," Geology gives us distinct evidence that God has acted in finite Time, and in local Space, as well as throughout the universe from eternity.

I maintain, therefore, that the evidences of Christianity, and the narratives of the Gospel, are to be judged by the laws of historical evidence, from consciousness and testimony; and that the Natural Sciences have nothing to do with the subject, except as they modify in other ways our History and Psychology.

I now return to a nearer view of that great division of human science which I call Natural History, the second division in the logical order of pursuit.

This department of Science takes in all the world of matter, and may be subdivided into three smaller departments, according as the forces of which it treats are mechanical, chemical, or vital.

I place Mechanics in the first place, because its dependence on mathematics is more close and definite, and because a knowledge of mechanical forces is necessary for understanding the action of chemical and vital powers. Under this head of Mechanics come Astronomy, Machinery, Heat, Light, Sound, and fluid action, all of which in nature take precedence of chemical and vital physics, because their laws are more general, and chemical and vital phenomena are dependent on them.

I place Chemistry next, because its dependence on Mechanics is closer, and because also vital phenomena depend upon chemical.

The third division of my grand department of Natural History treats of the phenomena of life. This is what is called Natural History proper, and is the department in which I am to suppose you more particularly interested. I wish, however, to repeat, that my remarks are not drawn from experience, but only from a general survey of the field.

It appears to me, then, that the successful study of the phenomena of living things requires some knowledge of Mathematics, Mechanics, and Chemistry; and that no degree of proficiency in any one of these preliminary branches of knowledge will be useless to the anatomist, botanist, or zoölogist. To select a few striking examples of the value of preliminary attainments, take the law of phyllotaxis in botany, and the form of the embryo in zoölogy, and you have two problems of vital phenomena that cannot possibly be solved without a respectable amount of mathematical skill. Take, again, the effect of centrifugal force on the radicle in germination, the force of ascending sap, the hybernation of animals, the distribution of plants according to climate, and you have problems that admit of no solution, unless by the aid of mechanics, the theory of fluids, and the facts of astronomy. Once more, take the comparative value of different kinds of food and manures, or the effect of galvanic and electric agencies, and you find a set of problems in which nothing but a profound knowledge of chemistry can avail you.

In the consideration of the phenomena of life, we may, I think, divide the subject into three heads. We may consider either the forms of organized beings, or their functions, or their habits.

The consideration of the forms of plants and animals is by Comte called Statical Biology, or, as we might express his idea, the statics of vital force. I prefer, however, in this branch of the subject, to ignore the forces of life. I think that in the anatomy of plants and animals, and in their classification, we are guided wholly by the shape, color, and size, by what may be called the form of the being, the geometry of organization.

The anatomy of organized beings, and their classification, logically precede the study of their functions and habits, being more directly dependent on Mathematics, and less upon Chemistry; being also necessary preliminaries to the knowledge of functions and habits. But of the two branches, Anatomy and Classification, it is, I think, more difficult to decide which takes precedence. The classification is founded upon the organization, and might therefore seem to succeed the study of anatomy. But, on the other hand, the organization varies according to the class, and therefore it might seem that a knowledge of the classification was necessary for the knowledge of the organization. Thus it appears that the two things ought to be pursued together.

Our knowledge of the form of a being is not perfect unless we know all its forms, from the earliest germ to the latest decay. No one knows a plant thoroughly unless he can recognize it in any stage of its growth; no one knows an animal thoroughly unless he can distinguish it in every age of its life. Thus embryology is a necessary part of anatomy and a necessary prelude to a perfect classification, whether in botany or zoölogy.

In Whewell's Philosophy of the Inductive Sciences, he supposes that we need, for classifying organized beings, the peculiar idea of a type. But I doubt very much if this be so. The species can certainly be definitely described by external marks. The difference of species is the unit of classification, and some naturalists have asserted that there is no other difference existing in nature; that is, that genera and families are merely arbitrary groups of man's invention. And if Whewell's idea of classifying by likeness to a type be correct, then I cannot see why classification is not open to this reproach. Moreover, if it be true that difference of species is the only difference in nature, then I see not why the genera and other more general divisions should not be arranged simply with reference to the convenience of learners. But classification is, in my opinion, a much nobler work than this. If species are grouped into genera according to sharply defined anatomic distinctions, and these again into families, and the families into classes, according to real differences of organization, then classification proceeds according to the law of creation, and gives us a real admission into the plans and purposes of the Creator. It then asserts loftily its place to stand among the very chief of sciences.

Stepping now into the second division of Natural History proper, we study the function of organs, and the genesis of disease. This field, essentially different from anatomy and classification, lies yet so contiguous, that there can be no proper knowledge of Physiology without a previous acquaintance with anatomy. And in point of fact the knowledge of functions is much less developed than the knowledge of forms. For although the cellular structure of organized beings, and the true divisions of genera and tribes, are as yet imperfectly known; yet on many points of physiology and nosology our ignorance is immeasurably more profound. Where is the man that can fathom the chemical secrets of the glands, and show by what nice manipulation the various secretions are formed? Who can explain the persistence of type in a disease, and show how small-pox virus produces in the human skin small-pox, and why that virus introduced into the skin of the ox produces an equally marked but distinct disease? Who will explain the specific and definite abnormal growth in a plant, producing a nidus for the progeny of the insect that pierced it?

I need but allude to such questions as these, to show what an unbounded field lies unexplored before the eye of the physiologist.

One fundamental question in physiology concerns the nature of the vital forces. Can they be reduced to the forces of heat, light, and electricity, or is there some peculiar principle of vitality which is distinct from these universal agents, and recognizes them only as assistants or servants? Whatever may be the final decision of Science upon this point, I think the evidence hitherto adduced is all in favor of supposing in simple organic growth a force entirely different from those which control the world of inorganic matter; and that the reality of vital powers should be made, to say the least, the leading hypothesis in physiology, until further evidence accumulate.

Comte defines the grand problem of physiology in this way: Any two of these three terms, an organ, the place, and the function, being given, to find the third.

An animal or a plant, or any part of either, this is the first term. The place or medium in which the plant or animal is situated, its temperature, degree of light, and moisture and air, in short, the surroundings of the creature, constitute the second term. And the third is the function, the mode of manifestation, and the product or effect. Of these three if two are given we are to find the third.

But truly this statement of the problem implies faith in more than observation can give. It implies a faith that the world is built in perfect harmony, and that each animal is a perfect unit in itself, and perfectly adapted to the place it fills.

Comte, however, denies this faith on which he builds, and asserts that the medium in which an animal or plant is placed determines its organization and its functions. And he extends this assertion beyond Physiology even into Psychology. The only argument which he gives on the subject is the following sentence. "The most obstinate psychologist would, doubtless, be unable to persist in maintaining the sovereign independence of his intellectual entities, if he would only deign to reflect, for example, that a simple momentary inversion of his ordinary vertical position would be enough to interpose, at once, an insurmountable obstacle to the course of his own speculations." This argument seems to me about on a par with Mr. Mill's pepper in the cook. No one denies or can deny that the human soul, if there be a soul within us, dwells in the human body, and that the human body is fitted to live only when certain external conditions are fulfilled. But to argue hence, that these external conditions create the soul, is as absurd as it would be to suppose that temperature, gravity, and atmospheric pressure are the motive powers of a mantel clock which stops if drowned under water, or if frozen by intense cold, or if subjected to a momentary inversion of its ordinary vertical position. None of these experiments disprove the theory that the mantel clock moves by the elasticity of a spring, and no dependence of organized beings on external circumstances shows the nonentity of vital forces and of intellectual realities.

I must now pass in my rapid survey to the third department of Natural History proper, and that is what I may call the Psychology of brutes. This department lies veiled in the greatest obscurity of the three. It depends very much on our own consciousness to interpret the facts which observation gives ; and it requires the nicest discrimination, lest we carry into the brute mind that which is not there, or exclude that which it really possesses.

But I cannot but be persuaded that the study of the habits of plants, as compared with those of animals, and more especially the comparison of the habits of animals with each other and with the habits of men, is destined hereafter to throw great light upon all questions of Psychology. As Comparative Anatomy is the only real and thorough Anatomy, so Comparative Psychology is the only real and thorough means of acquiring a knowledge of the soul.

I regret that the length of my previous remarks has left me little time for the development of this part of the subject; as it was that of which I wished more particularly to speak. I conceive that we have here latent an instrument of vast power for the true development of mental philosophy. I hold with Professor Bowen, that one experiment on congenital cataract has done more for the confirmation of the metaphysics of sensation than can be well estimated. In like manner, I believe that intelligent and careful observation of idiotic, insane, and eccentric people, and the comparison of their mental processes with our own, would do much for the advance of the true knowledge of our mental powers. Of equal or greater value will be found careful observation of the development of intelligence in the child. We should study, if I may so phrase it, the embryology of the mind. If Science did not demand this, justice to our children would, for they are very often unjustly judged, by treating them as though they were adults. But Science does demand it; — we may gain the most valuable knowledge of psychological analysis from watching the development of mind in a child even within an hour after birth.

I think that similar advantages, though not so great, may be gained from the examination of the reasonings and instincts of animals; a subject on which almost every naturalist has generalized far too hastily, and with too much subservience to his preconceived ideas.

But I have trespassed already too far upon your patience, and will close by explaining my object in this dry and general analysis of Natural History and its relation to other sciences. I did not hope to give you any aid in the detail and methods of your pursuits, except so far as might be done by pointing out to you what I conceived to be the true objects of Natural History, and its relation to other pursuits. I wished you to see that in the natural hierarchy of Science, in the order in which they must be studied, Mathematics and Chemistry precede Natural History, and Natural History precedes Psychology and Theology. In one sense the Mathematics are the servants of all sciences, and all sciences servants of Theology. But in a juster view all the five grand departments are equally noble, as each makes us acquainted with the works of God, and with his thoughts, and develops in us those powers which constitute that spiritual likeness to him in which we were created.

If you study Natural History in this generous spirit, devoting yourselves to Physiology and to the study of the habits and instincts of beings, as well as of their anatomy and classification, you will then redeem the science from the reproach of being, at least in your hands, a barren catalogue of Latin names.

Thanking you for your kind attention, I have done.

www.ingramcontent.com/pod-product-compliance
Lightning Source LLC
LaVergne TN
LVHW011139110826
845150LV00008B/2413

9781418193737